THE ELUSIVE TEN

▼

A New Deal in Revolver Target-Shooting

By

WILLIAM REICHENBACH

Originally published in 1935

ISBN: 978-1-940001-22-7

Published by Sportsman's Vintage Press
2014

Dedicated to my wife, Agnes Reichenbach

AN APOLOGY

Due to the Printer's oversight there are several mistakes in the printing. We apologize.

Page 20—Last line should read:
"Bottom: Colt Officer's."

Page 32—Line 16 should read:
"can safely relegate most of the wonder stories to the"

Page 59—Line 17 should read:
"elusive. This fact, in addition to being very"

Page 71—Insert between lines 17 and 18 the following:
"made method of autosuggestion? Could the"

CONTENTS

MY ALIBI

I believe it is generally conceded that every normal human being, sometime in the course of his life, wishes to write a book. This desire is indulgently accepted by his social circle as a form of mild insanity, similar to the urge of writing love letters or, worse yet, soulful poems, without the attendant secrecy. It is regarded as something inevitable. The fact that a certain technique is required to make the product less objectionable, does not always bother the perpetrator. The effort which may be likened to "letting off steam", forms a sort of safety valve for almost anything. But sometimes, the patient really thinks that he has something to say that the world should not do without.

I wish to assure my readers that although I have long felt the inspiration for book-writing, I have resisted the urge throughout the years, doubtless to the greater satisfaction of my many good friends. However, the ways of fate are strange. One irresponsible companion induced me, after many years of idleness, to take up revolver-shooting again. I fell to it with zest. Being sort of thoroughgoing, I shot and studied at the same time. Although I shot well, I was not happy. The fact that many of my associates did not seem to catch on to the art was extremely disturbing to me. They assured me that they had studied all available literature on the subject. That they were trying, painstakingly and untiringly, was obvious. I therefore proceeded to give them some pointers. And lo and behold! (what it means I don't know, but lo and behold!) they progressed.

And an idea was born.

Maybe there are **other** shooters who have tried and have never succeeded? Maybe I really have a message to convey? Maybe I have stumbled onto something too valuable to be allowed to perish in obscurity? Maybe I can satisfy a demand for an illuminating treatise? Try and stop me.

Seriously though, the cause for the poor performances generally seen in revolver shooting, is to be found in inadequate knowledge of the true technique. Coaching, with rare exceptions, is done along antiquated lines, wholly lacking in results. My studies have convinced me that only a complete departure from the trodden paths of outdated teaching would provide the solution to the difficulty.

And if I had been contented with this discovery, this Manual would never have seen the light of day.

But the trouble with me is that I am so serious about this thing that I cannot keep it for myself. I feel that I **must** relate my own experience so that others may benefit. If you suffer my book in silence, and if I cannot stir you out of your lethargy and force you to take the matter just as seriously as I do, then I must blame my writing technique. The matter itself, my dear and hopeful reader, is worth any one's while. The question, therefore, seems to be less one of justification for the conception of this book than one of finding apologies for its style, or lack of style. Although old-timers may shoot me on sight, I gamble on being useful to you, my dear reader. You alone are the judge to say whether I have the right to burst out in writing.

Assuming that I shall be forgiven, I proceed to call my brain-child a MANUAL. It probably is one. It sounds rather nice. To start with, I have made a firm resolution to let this Manual deal solely with the technique of holding, aiming and firing a revolver. If you seek information on internal and external ballistics too, then I advise you to ask one of the guncranks down the block. They are more numerous than wild rabbits. The bit of technical stuff which you may find in my Manual was just thrown in to impress you.

As for the title, I settled on "THE ELUSIVE TEN, a New Deal in Revolver Target-Shooting". (The "Ten" referring, of course, to the "Ten-Ring" of the target.) But why should the "Ten" be described as elusive? If you do not already know the answer, you soon will.

Satisfaction at having realized a latent propensity for writing, is complete only when the driving factor has been the honest urge to submit knowledge gained in study, knowledge which we know is lacking in many quarters. It seems simply unbearable to know that with the excellent material we have in our country, we should have so few successful revolver shots. The irony of it! The revolver was invented and perfected right here and it actually played an important role in the building of our vast empire and yet we are essentially not a nation of revolver shots. Europe has been much more consistent. There the modern automatic was conceived and the actual percentage of real marksmen over there brings home, quite forcibly, how really sad the picture looks in our country. What is the cause of that? It cannot be blundering legislation only, controlling the distribution of hand-arms, because Europe too has legislation of various kinds. I believe, it is nothing but lethargy, pure and simple. Indifference toward a sport which imposes so much control over one's self. Are we Americans really steeped so much in nervous haste, in impatient tackling of vivacious sports only, sports which demand feverish action? Football, Baseball, Boxing, etc.? Not quite. Take fishing, or its more scientific branch: Bait Casting. We have there many thousands of really good performers. Why should that sport which surely demands skill, patience, knowledge, nerve control to a high degree, produce so many experts? The answer is simple: The equipment and the technique have been developed methodically and logically. Cannot we say this about revolver shooting also? I am afraid that the answer must be partly in the negative. Without a logically developed technique we discourage too many aspirants. It appears as if only sporadic efforts have ever been made on the part of those that should have been leaders, namely the experts in our sport. It is true that we have had and still have some remarkable shots, but they either tackled the job of distributing their knowledge in quite an unsuitable or dispirited way, or not at all.

I have made an effort in this Manual to fill the gap and do hope that interest will be stirred up among the coming experts to cause them to take a leading role in the development and the distribution of the true technique of revolver shooting. Since we in America still cling to the revolver, as witness our vast police forces, guards, most clubs, etc., who all favor the revolver, how important, how vitally important, then is it to provide them with "technique", which up to now they do not possess.

I venture the prediction that, since this Manual contains so much controversial matter, there will be criticism galore. That would be fine. Anything is better than the present indifference on the part of our should-be leaders.

We need new blood —

New ideas —

A New Deal.

PART I

VENERABLES OF THE OLYMP

Webster will concur that I imply no slur.

My assertion that I choose the above title for our Old-Timers in the game with a feeling of regretful futility should not be significant. Confidentially though, I am reminded of the gentleman who politely suggested to a dear (?) pest to take the elevator (adding, under his breath, not to trouble to wait for the car).

Maestro, dear Old-Timer, where, oh where, is the crop of successful shooters which your ability should have developed?

I wish to say here that I admire nothing more than an expert revolver shot, and, naturally, my deep admiration includes you Old-Timers. However, I must ask now how much of your experience have you divulged, you rascals? Were not most of you satisfied to play cock of the walk, soaking up the dumb admiration of your colleagues? — Facile princeps.

It is true that some of you have written books, but because of their high prices, they were inaccessible to the army of revolver enthusiasts. Some of you have approached the matter in a scientific way, which is so much Greek to us dubs.

Were we beginners not left more or less to our own devices?

What we **wanted** and **needed** were precise instructions available to every one — directions guiding us step by step, not the agony of having to find out things by our lone selves, by the trial and error method. How else were it possible that we still have so many thousands of enthusiasts who have never achieved even the mediocre?

Take the vast number of army men, police officers, guards, etc., the majority of whom handle their guns in such a desultory fashion, that, to break their necks, they cannot even hit the proverbial barndoor.

Will my little book be a good mentor?

I aim to address only men and women of intelligence, thereby reducing the number of possible failures by a great percentage. I believe that the group which proves to be intelligent enough to digest my treatise will get somewhere. They may not all turn out to be champions, but they will give good accounts of themselves.

I plan to use a suggestive method, cloaked in conversational language, hoping it will prove to be helpful. It has been, where I have instructed personally. The written word, naturally, cannot fully replace actual demonstration. We all know that. But, for intelligent people, great elaborations or endless repetitions will not be necessary.

Although I trust that I am offering a concise system of my own, there is no issue at stake between you Old-Timers and me. I ask no indulgence from you. Things such as preponderance of **grip** over **squeeze**, or vice versa, are only matters of opinion and should not be subjects of dispute.

I could point out the fact that systems, any systems, are debatable, but their only criterion must be their success or their failure.

And, my dear Old-Timers, we do really have

too many, entirely too many dubs — shall I say — in spite of you?

BY WAY OF GETTING ACQUAINTED

To you of my readers who are new to our sport of revolver-target-shooting, a few words:—

Have not all of us, already as a child, heard or read of wonderful feats of marksmanship? Of duels among the gentlemen of the South? Of the deadly accuracy of our Western gunmen, peace-officer and outlaw? Have we not with bated breath followed the exploits of our heros and how often and fervently have we not wished to command the same daring and unfailing marksmanship with a six-shooter? Even now, as grown-ups, the romance of those glamorous adventures touches a receptive chord in our minds. We may have become a little more practical, a little less easily moved, a little skeptical, but we still cherish deep in our heart a desire to be a deadly marksman with a six-gun and to be known as such.

But we dismiss the thought as just a little foolish. What would be the use? And, anyhow, it would be too difficult.

Dear readers, young and old, why should such a desire be foolish? True, we would not want to re-instate the rather wild and lawless times of the gold rushes, of the border towns, of Indian warfare. But, why should we not indulge in revolver shooting as a **sport**? Do you know that many, many thousands of us have been shooting with a six-shooter for years? That there existed all along a fully organized sport of target-shooting? Well, it is the truth. Every year, as the sport becomes more popular, many thousands more become interested and are converted into Target-Addicts.

And one of the objects of this Manual is to make our sport more popular.

Still, you believe, that it would be too difficult to take up. Let me tell you, and I really know what I am talking about, practically anybody can become an expert at our game. You too.

It is admitted that so far, the systems of teaching have been rather desultory. It was more or less a matter of one shooter giving the other little hints here and there. Although there has been altogether too much mystery around good marksmanship, the sport has grown by leaps and bounds.

But, there need be no mystery any more. This Manual has been written to initiate you along easy and logical lines. Read it and get interested.

Our sport, by the way, is not expensive at all, but what you get out of it, will be priceless. Your whole bearing will exude confidence. Your nervous system will improve. Your attitude toward the world in general will become more tolerant, with the knowledge that the richest man in the world cannot buy, with all his money and power, your ability to shoot well with a six-gun.

Study the Manual carefully and practice as outlined and I promise you that, within a very short time, you will be able to astonish your friends with your good marksmanship and thus realize a long cherished dream.

(And you get something for nothing, namely mastery of the rifle. A good revolver shot will **always** shoot a rifle equally well.)

Revolver-shooting is, in my opinion, unquestionably the finest of all sports. This may impress you as a rather broad statement and I feel that it requires substantiation of a sort (even if I should run the danger of possible expostulations on the part of my readers — you know: general objectionability to personal approach, etc., etc.)

I was a good fencer and a good oarsman; in fact, I was an active and ardent follower of many sports. But none have ever appealed to me as much as revolver-shooting. My interest has never been so constant, so intense. I have never experienced the thrilling moments of complete satisfaction in any sport that I have experienced when making an exceptionally good target. Why is this? Because I am a fairly successful revolver shot? No, I think not. There is a deeper reason, although I admit that one's liking for a sport or pastime is much dependent upon the extent of his success. Let me explain.

Consider first those who comprise the membership of revolver clubs—our associates. At first glance, one is apt to think that they could not possibly form a harmonious group because of the many variations in daily interest demanded by their professions or business. Nevertheless, among them we find a generous and helpful spirit of cooperation, of mutual respect, and of decent tolerance. This is not surprising, for the revolver and target draw only the best—those of high intelligence and unquestioned integrity. Indeed, the four-flusher, the chiseler, and kindred types are conspicuous by their absence from our ranks. As a class they can see nothing in it. They are helpless against the candid honesty of the firing line. In short, revolver-shooting is a gentleman's sport.

But there are other intriguing aspects. Give a thought to the target revolver and to him who shoots it. It is hardly conceivable that any one would take up our sport without an inherent liking for firearms. True, he may know little at first, but as he continues he cannot help but learn much. He may change from one type of gun to another, but eventually he will find "his gun," and thenceforth, a partnership is in the making. He may, on the other hand, with stubborn insistency set out to conquer one model and strike a lifelong friendship that way. He learns about its proper care, he learns to respect its honesty, to put his faith in it. And in return, his partner scores for him exactly what he deserves—no more, no less. With the passing of time he develops a regard for his favorite weapon that is nothing short of affection. Such is the partnership between a man and his gun.

Practically all sports require the use of vibrant muscular action. Some require more, some less. But, with revolver-shooting the opposite obtains. Our successes, if we are to have them, demand controlled **inaction** — a condition closely approaching utter and complete muscular relaxation. And we remain relaxed throughout the ordeal of firing. There is no physical outlet for pent-up emotions—no golf ball to drive a couple hundred yards, no tennis ball to serve with every ounce of our strength. Although our nerves may be jumping and our hearts pounding from the excitement of competition, we must strive to get control and so remain. Easy? It looks as though it might be. But, can we do it? Are we our own masters? What a personal challenge!

There you are. All this is what makes our sport well worthy of the highest type man or woman. Whether you work with your brain or hands, whether you are a great industrial leader or a private in the ranks, a cordial welcome awaits you.

CONCERNING YOU AND ME

Just how serious are you about revolver-shooting?

I assume either that you know very little about it and desire to learn, or that you have done some shooting and are eager to improve your scores. But a simple casual desire is not enough. The target revolver demands much more of those who really want to master it.

Do you play golf? What is your handicap? Why is it not lower? Can you improve your game? How? Then, why don't you?

Do you fish? Are you a **good** fisherman, invariably proud of your catch? Have you the patience, the knowledge, and the skill which that sport requires? Do you really know **much** about it? Have you ever missed an opportunity to learn more?

Do you play bridge? Ashamed of your game? Do you spend most of the time apologizing? Or have you always endeavored to improve your play by seeking out better players than yourself so that you could learn from them?

Consider any sport or pastime with which you are familiar. Ask yourself questions of a similar nature. Be frank with your answers. Place the blame for mediocrity where it belongs. Don't fool yourself.

To succeed in revolver shooting (and by that I mean to score consistently well above the average) will require something of you. Unlimited patience, constructive self-analysis, and non-flagging enthusiasm are some of the absolute essentials. You must learn to wage a winning battle against the perils of discouragement. You must deny yourself the luxury of emotional outbursts of temper. Indeed, the matter of consistent high scoring is dependent on far more than just the eye, the hand, and the trigger finger. You must go behind the obvious exterior, to the nervous system, the centre of your physical control. Before you start, you must gauge the extent of the importance of control. Approach the sport with the determination that your foundation shall rest on solid, substantial, bed-rock **control** rather than on the treacherous sands of shifting and excitable emotion. In short: Keep cool. Be willing to try again. And always, always keep your chin up!

One more point. You must have a real, intense desire to improve constantly. You must find yourself looking forward with enthusiastic anticipation to the next time your club shoots. You must have that indefinable urge to do better. And when you don't do better, as will sometimes be the case (for, after all, this is not an easy sport), do not permit that overwhelming urge to lessen. A casual interest cannot carry you to the heights.

Can I help you? Yes, if you will but put yourself in my hands. By that I mean: be absolutely willing to accept certain conditions, prerequisite to our mutual success — I with you and you with the target revolver. Second: digest each lesson thoroughly before taking up the next in sequence. Third: understand that this manual is not composed of a collection of helpful hints; on the contrary, it is a definite system of instruction, intended to develop you ultimately into a competent revolver shot.

And may I ask you the same question a second time? Just how serious are you about revolver-shooting? At least you should know by now what the answer should be.

HOW MUCH IS THERE TO LEARN?

No doubt, you have heard or read about some individual popularly known as a "born crack shot." Disabuse your mind of the thought that any one is so fortunate. Crack shots are not born; they are made — only as a result of hard work. Some arrive easier and go farther, because of excellent temperament and physical characteristics of a type ideal for the target revolver. But that is true in all sports. The "natural" is fairly common, particularly in track. But I seriously question whether a "revolver natural" possesses very much advantage over his fellow neophytes. In fact, I assert that, barring physical impairments, any one of reasonable intelligence can be developed into a competent revolver shot — yes, into a crack shot — **if he will but understand and work hard!**

By work I do not so much mean endless practice, but rather constant and intelligent thought. If you start right, both mentally and physically, you cannot fail to do well. But starting right means starting at the beginning — not at the end. Before you can consider yourself competent to sight your weapon on a target and squeeze the trigger—even though the gun be empty—there is much to learn.

In the following chapters of Part I which deal with elementary fundamentals, you will note that a definite order of sequence obtains. Follow it. Do not skip about looking for hints. Again let me say, that is not the subject of this Manual. Adhere to the course with thoroughness and your gradual development on a sound basis will be assured.

Moreover, the arrangement is most logical. First, you learn how to grip your revolver. Then, how to stand. Third, how to relax—how to acquire a feeling of nonchalance which will inspire confidence. Fourth, how to start the gun from shoulder toward the target and to continue until the arm is extended and you are in a position to sight it.

(All of that—and you have not yet reached the usual thing every one does immediately upon first picking up a revolver—sighting it and snapping the trigger.)

Pause for a moment, if you will, and believe that there is much to learn, much to master before you even attempt to sight your gun.

We then take up sighting. But even now you are not ready. We must determine which is your master eye, the one you must use for your "bead." You will be told how to sight and where to "hold" on the target.

Finally, we come to squeezing the trigger and firing the shot, with a consistent follow-through of the trigger finger as the revolver is brought back to the body, and we are made to realize that a definite breathing technique is required.

Seven distinct fundamentals you must keep uppermost in your mind until they become second nature with you. (Grip, Stance, Relaxation, Bringing the gun forward, Sighting, Squeeze and Breathing). And when they do that very thing your scores on the Standard American Slow-fire Target at 20 yards should rarely go below 80.

Still interested? Then promise yourself that you will adhere strictly to the instructions in the following chapters and that you will regularly follow the suggested practice exercise. Do these things, and you will get along.

GUNS

As a logical opening chapter to the practical part of the manual, I offer to my readers in general, but particularly to those who have not yet bought their guns, a short cross-section of suitable target-weapons.

Keeping in mind the fact that this Manual is devoted primarily to the American revolver, I feel that mention should be made of at least one automatic. Lack of space forces me to omit some suitable revolver models (Harrington and Richardson, for instance, brought out some very creditable 22-calibre weapons at nominal prices), Although I do so with some regret.

There is hardly a piece of sportive equipment which meets with such jealous partiality as does the target gun. However, I know better than to let myself be drawn into endless arguments **pro and con,** and I shall confine myself to the statement that the fine target weapons brought out by the leading American manufacturers all have their merits. In the final analysis, it is therefore up to you to pick your own gun.

You will notice that I show two kinds of calibres — namely the 22-calibre, which fires a rim-fire shell, and the 38-cal. using a centre-fire cartridge.

In taking your choice, you should keep in mind what kind of work or competition you intend to take up. In many competitions, for instance, rim-fire weapons are not permissible against centre-fire guns, the contention being that the heavier recoil of the latter would impose an undue disadvantage.

The rim-fire shells are very economical, and in view of the lack of noticeable recoil and muzzle blast, this calibre is a great favorite. The centre-fire 38-calibre revolver, on the other hand, is a real he-man's weapon, and although a little harder to conquer, many experts swear by it and look with disdain upon the smaller calibres.

In a way, it is a toss-up which calibre would give better results in target work. While the smaller calibres are easier to handle, the bullet holes made by the larger calibres cover more territory.

There is another category of target guns which I shall, intentionally, leave out. I refer to the so-called "Free Pistols." These are very extreme single-shot models, on which science and ingenuity have expended their help to an almost uncanny degree. Short of leather-upholstered resting facilities and melodious musical accompaniment, they are the last word in hand guns. They are, however, frisky and unwieldy, besides being very expensive, and for our purpose we can forget them without regret. They are barred from most matches.

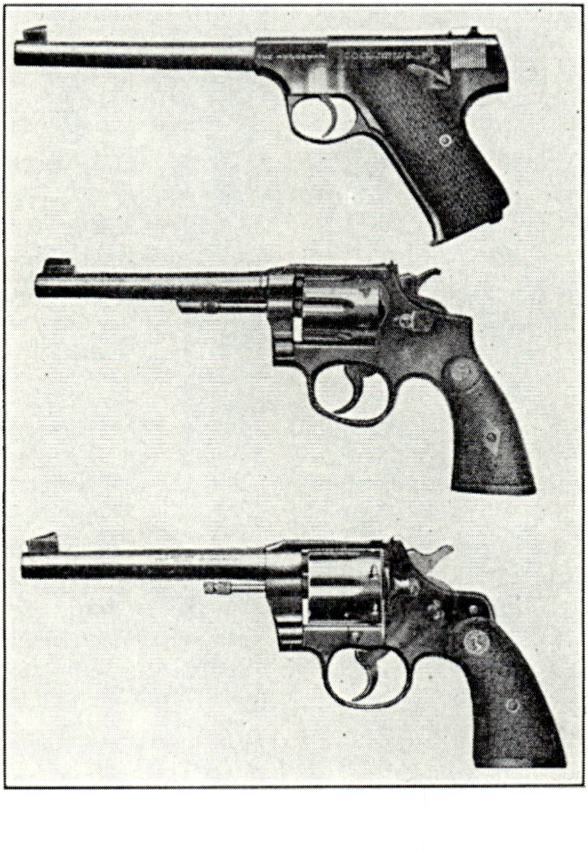

Top: Colt Woodsman.

Centre: Smith and Wesson, Model K.

Botom: Colt Officer's.

SMITH & WESSON
6 Shot Revolver,
Model K.
22 cal. Rim-Fire

Very smooth action and an excellent target gun.

SMITH & WESSON
Revolver, Military & Police, 38 cal. Centre-fire, 6" barrel. (The larger sister to the 22-cal. Model K)

It has the same agreeable action as the K22.

COLT WOODSMAN
Rim-fire Automatic
22-cal. 6" barrel

A very useful hand-weapon, not confined to target work alone. It is remarkably accurate and handsome.

COLT OFFICER'S MODEL, 6-Shot Revolver, 22 - cal. - Rim-fire, 6" barrel

If ordered from the factory with hand-honed action and a specified trigger pull, this weapon is unexcelled and the primary choice of many target shooters.

COLT OFFICER'S MODEL, 38 cal.-Centre fire, 6" barrel (The pendant to the 22 cal. Officer's model).

This is unquestionably the outstanding target gun in this calibre and eminently suitable for very close shooting.

ILLUSTRATION I.

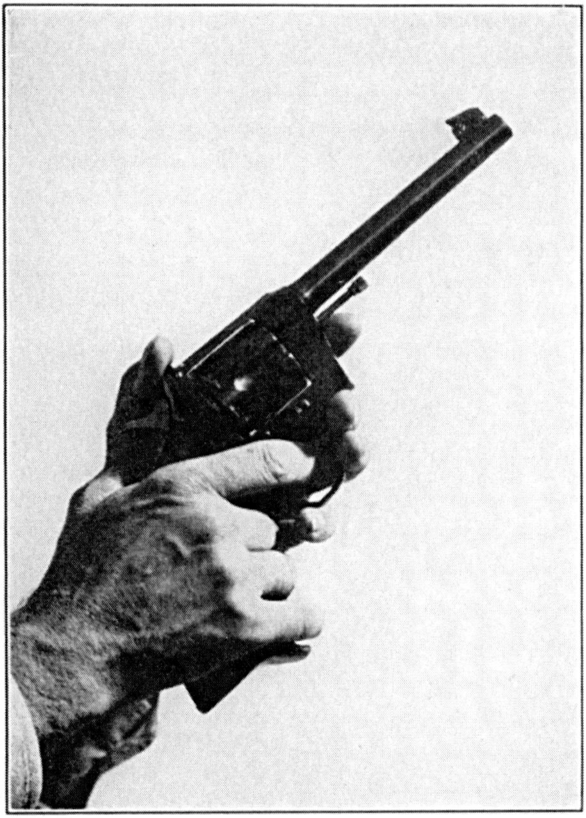

FITTING THE GUN

The five fingers of the left hand support the weight of the gun until the fitting into the right hand is completed.

If the gun should be cocked the left thumb which rests under the hammer, is withdrawn only after the fitting is completed.

GRIP

Considerable emphasis should be placed on proper grip. While you, being an intelligent person, would not attempt to hold your gun as if it were a coffee spoon or a tomahawk, you will endeavor, at any rate, to take some natural grip which will cause the least fatigue. Something like this:

> Hold the stock of your gun between the ball of your thumb and your middlefinger. (You really could shoot that way.) But, in order to put the other fingers away, you might just as well place your thumb along the upper edge of the plate (on the cylinder latch), parallel to the hammer, and
>
> Drape the ring-and little fingers gracefully wherever they find a place on the grip of the gun.
>
> Mind you, just drape the fingers around the stock, and **do not touch the stock with your fingertips at all.**

The lower part of the trigger finger touches the frame and should steady it. The little finger, also with its lower portion, does the same further down the stock.

The action of the first two digits of the trigger-finger is being dealt with under the chapter entitled Squeeze.

That seems to be the whole secret of Grip. However, it may not be amiss to amplify this statement. There is a reason for everything, as you will see presently.

Let us go at it systematically.

Hold on! Do not pick up the revolver with your right hand and manipulate it into position by the fingers of that hand unaided by the left. No, sir! Instead, hold the weapon by the left hand and "fit" it into your right hand (See Illustration I). Never overlook an opportunity to give your right hand every possible help, no matter how inconsequential it may seem.

ILLUSTRATION II

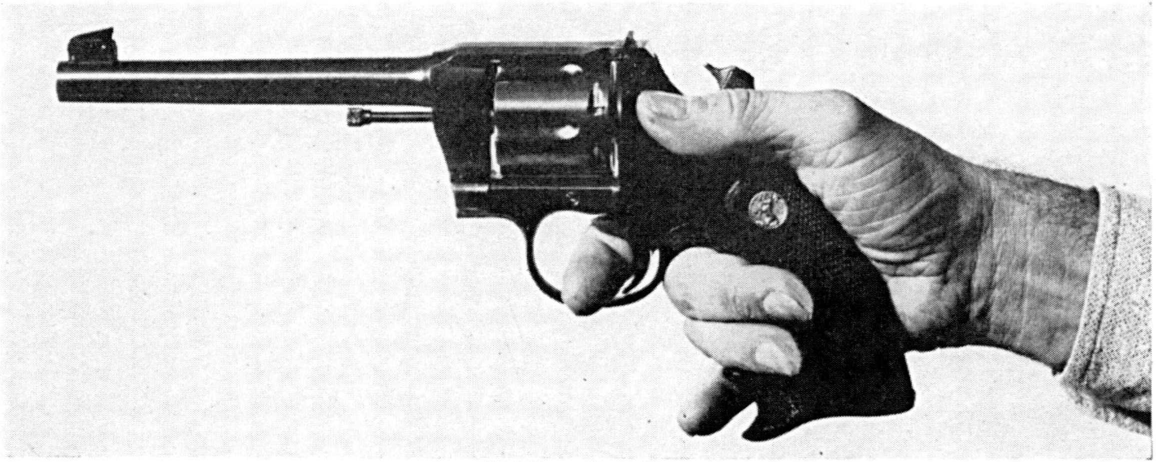

(Holding the Gun)

The general impression which the picture gives is that of complete absence of strain or effort. It appears as if the hand were holding a musical instrument.

Observe the shadows which middle- and ring-fingers throw on the stock of the gun, illustrating that the finger-tips are kept away from the stock altogether. The thumb rests on the cylinder-latch. One can distinctly see the little-finger support the right side of the stock while the tip is also kept away from the stock.

How the revolver should be grasped is illustrated by the accompanying illustration (See Illustration II). Note particularly the position of thumb and fingertips.

Although the revolver handle is known as its "grip," that should be no reason for gripping it as though you feared some one might attempt to yank it out of your hand.

A short explanation of the various forces involved in grasping a revolver properly should prove helpful, not only as a matter of correct learning, but also as an aid in acquiring consistency. The weapon is supported by the middle finger; in fact, it literally hangs on it. Close to the second joint and on the side of the middle finger (i.e. the side nearer the thumb) is the point of support. On the revolver, the corresponding point is directly in the rear of the trigger guard. This point serves also as a fulcrum. The pressure exerted by the middle, ring, and little fingers towards the rear is opposed by the ball of the thumb. The vertical movement of the muzzle, caused by the working of the mechanism and recoil, is checked by the thumb resting on the cylinder latch and by the ball of the thumb The horizontal movement of the muzzle is governed by the palm and the sides of the trigger-and little fingers, as mentioned before and again the thumb, by its strategic position on the cylinder latch, comes into passive function.

Understand the foregoing and accept the fact that, contrary to the beginner's usual conception, the revolver may be grasped almost lightly — in fact, just firmly enough to withstand the shock of recoil. That, you will quickly find, is beneficial in many ways. It is less tiring. The gun remains steadier. Your motions and reactions are less accute, etc. Just regard the revolver as a delicate instrument and hold it accordingly. (See Illustration II.)

By now you will readily appreciate the importance of what I said previously. **Do not touch the revolver grip with your finger tips!** The slight pressure they exert, the amount of which you cannot control, will destroy the delicate state of balanced forces (opposing pressures) which you are seeking to establish and maintain until after the weapon is discharged.

Practice holding your revolver. Study the illustration II. Observe how the various forces (pressures) which you exert, balance and offset each other. Do not be satisfied until the revolver feels natural and until you grasp it in exactly the same manner each time you take hold of it.

STANCE

"Stance" is a beautiful word, and it is also a beautiful thing to look at, if taken properly. It means "way of standing". Without a good stance, one not only looks bad, but one will never attain the heights. One would think that nobody would wish to look awkward anyhow — surely no one with intelligence would like that.

If you have been shooting with a wrong position, just forget everything you knew or thought you knew up till now. Use your will power to forget old habits which are standing in your way to progress.

It is unfortunate that too often the beginner pays little or no attention to the matter of stance. And, knowing nothing about its effect on his scoring, because of body sway, he handicaps himself unwittingly. Yet, it is all exceedingly simple.

First, we have to realize that no matter how we stand or place our two feet, there will be a more or less noticeable sway of the body. If nature had provided us with three legs, or a sort of tripod as it were, body sway would not concern us.

If you were to stand with your feet apart, say 10 to 12" or thereabout, with your toes pointing to the left of the range, your right shoulder facing and right gun arm aiming in the direction of the target (See Illustration III), you would find that the muzzle of your gun would waive mostly from left to right and vice versa, because your body is swaying that way, that is across the line of firing.

Now, let your toes point toward the target and also present your chest to the target. The muzzle will now have the tendency to make longer travels in an up and down direction, because your body sways toward and away from the target, lowering and lifting your gun hand. (See Illustration IV).

Well, the body sway is an acknowledged fact and I mentioned it because it concerns us vitally.

What effect then has body sway on the grouping of shots? If you sway in the direction of the line of fire, they may spread vertically. Swaying across the line of fire causes a horizontal spread.

Obviously, the theoretically correct stance is somewhere between the two extremes mentioned, with the heels on a line at an angle of about 45 degrees with the line of fire. (See Illustration V) For consistency, always assume your stance by the following routine:

1. Stand with feet together facing in a direction of about 45 degrees to the left of the line of fire.
2. Move left foot straight to left, a distance of 10 to 14 inches, more or less, to find a comfortable position.

ILLUSTRATION III.

(Wrong position)

The body will sway toward and away from the reader. Consequently, the gun muzzle will oscillate horizontally across the target.

ILLUSTRATION IV.

(Wrong position)

The sway of the body is toward and away from the target and will cause the muzzle to bob up and down, vertically across the target.

ILLUSTRATION V.

(Correct position)

The sway of the body is toward and away from the lower left corner of the picture. The muzzle will waive vertically, horizontally and diagonally across the target, but the travels either way are greatly diminished, in contrast to those caused by the wrong stances shown in illustrations III and IV.

Although the shooter is absorbed in serious concentration on his task at hand, the whole body, from the little toe to the gun muzzle, suggests complete relaxation and bodily ease.

3. Now raise the heel of your right foot, turn on the ball of your foot to the left until you reach a position of comfort.

In the position now assumed the body, of course, will still sway, namely in the direction which your chest is facing.

However, you will not find your target there. Your target is more to the right and in order to aim at it, your gun arm has to be swung to the right, in a line between your chest and your shoulder. The muzzle will now not only sway diagonally, but also up and down and from left to right and vice versa, but the travel in either direction has been greatly shortened by the half twist in your shoulder position.

Any stance one can take is necessarily a compromise. Although you will sway slightly in every direction, even that may be diminished by placing your body weight correctly on both feet. Toe-in. Toe-in until you can feel pressure on the outer margin of each foot. Also, distribute the weight equally on both feet and on each foot evenly between the ball and the outer edge and the heel. We can see now why a person with three feet would have very little trouble at this stage, other than to untangle the members when he is ready to go home.

Remember, you are seeking the most solid foundation for your body that two feet can make. Keep that in mind. Approach the problem in this manner, modify it if necessary to suit your individual physical characteristics. But think about it. Don't dismiss it as trivial. And practice until you have developed a stance which is both comfortable and theoretically correct. Then practice some more for a few moments each day until you invariably assume the same stance each time attempted.

Shoulders, arms, body, legs and feet should not show any tension. As for the left hand, put it in your pocket and forget it. (Illustration V.)

RELAXATION

You have one of the most deadly and manly weapons in your hand that science has developed; yet the handling of it is to be of the most delicate nature imaginable. Brute strength and ferocious desire are only handicaps.

(It has always been a mystery to me how some of the gunmen of the frontier days could have been such marvelous shots, and I think we can safely relegate most the wonder stories to the ashcan by saying that we are satisfied to become only outstanding target shots.)

Relaxation —— Just what I said.

Forget that there may be anything martial about our sport.

Consider the thing in the light of a game of nerve against muscle, intelligence against brawn.

Take the tenseness entirely out of mind and body.

Just relax —

Relax completely —

No muscle need be tensed —

(Shooting does not require muscular exertion, and it would be assinine to tense your muscles when there is no call for it) —

You are not boxing or fencing —

You are just relaxing —

Verily a sport for lazybones.

You might also look up Webster again for a definition of "nonchalance." It will penetrate better that way.

Yes, we need a certain amount of nonchalance.

Do not try too hard. Take it easy. Do not be too much concerned.

While we do not wish to waste a single cartridge, and shall try to get every shot out along proved lines, we shall not act as though our very lives depended upon a perfect target.

True enough, we are serious and pedantic while we are shooting, but we should not conduct ourselves as though we had a large bet on the shot.

Nonchalance and complacence — very good requisites for a shooter.

We are not too extremely serious, but we certainly have quit fooling.

Either we shoot or we don't.

If you should have among your club members or your shooting friends some one who likes to monkey around, and, through his fooling or kidding, is apt to distract your mind from the concentration, which shooting surely demands, you will have to act.

As the well-known mule-driver said to his balky charge: "I shall either have to steel my nerves sufficiently so as to be able to overlook your antics, Aphrodite, or I shall have to !!*/!!!"

BRINGING THE GUN FORWARD

Now, my friends, we come to a chapter which may give the impression of having very little importance. Yet, I assure you, it is very essential. I think a good definition of what I have in mind would be:

When bringing the gun forward into shooting position try to describe with your hand a harmonious oval in an easy flowing, continuous movement.

The intent is more in the nature of an auto-suggestion, namely through an easy continuous flow of movement to abolish any and all jerkiness.

Jerks, abrupt and angular movements, are detrimental to the creation of the perfect atmosphere, in which all fibres in the whole body, through absolute relaxation (with the exception of one little part of the trigger finger), work together in ideal coordination.

To be explicit — with the cocked gun in your hand pointed upward and forward at an angle of 45 degrees, start a movement from your shoulder in the direction of the target (describing a **flat** arc) — aim — squeeze — **hold** the trigger — continue the arc flatly down and backwards — back to your body — **release** the trigger at the **end** of the completed oval — relax.

Again. Cock your gun. Start again from the shoulder (with the muzzle forward) — describe the first and upper part of the oval, in the direction of the target — aim — squeeze — **hold** trigger down — complete arc back to your body — **release** trigger — relax.

And the same with every shot.

This constant arc-movement, comprising, as it does, the follow-through (also found in golfing parlance), by keeping the trigger compressed after the shoot, refers to Slow Fire only. In Time- and Rapid Fire the gun is kept forward.

A deeply hidden precautionary measure is also put into practice. Never is the gun pointed carelessly at objects which are not meant to be aimed at.

SIGHTING

The mechanics of sighting are really not at all complicated, although sighting itself is still regarded by some persons as something of a mystery. The task is to line up the top of the rearsight-slot wih the top of the front sight and then with the target. The front sight should be centered in the slot of the rear sight (not too high or too low, but centered exactly) (see Illustration VI) and an imaginary line drawn from your shooting eye over the center of the rear sight slot and over the center of the front sight and leading to the 6 o'clock point on the target, would give the perfect "bead". (IllustrationVII.)

Furthermore, the gun, viewed from the rear, should be held so that the top of the rear sight appears perfectly horizontal or level. Do not tilt or "cant" the gun. (See Illustration VI.)

The sights shown in the illustrations VI and VII were cut from card board and then photographed. They represent square forms which we call "Patridge Sights". There are other shapes too, but, principally, they all serve the same purpose of guiding your sighting line over the barrel to the target.

Don't bend your head for sighting but rather bring the gun hand up high enough so that you may aim with a natural position of your head. (See also Illustration V.)

I am advocating in this manual to keep both eyes open when taking aim in order to permit a **plastic** and binocular view of both target and sights.

There is one thing at least to be said for this method — namely, that you do not have to screw up your face in order to close one eye.

Apart from the interrupted relaxation, closing one eye is a separate and positive muscular action which will detract from the intense concentration which should be directed solely towards shooting.

So, why screw up your face?

Have you ever tried aiming with both eyes open? It is really fun.

First, we have to discover which of your eyes does the work of lining up the sights and target. This working eye has been conveniently termed the Master Eye. It differs with individuals.

We proceed as follows:

Hold a pencil at an arm's length from eyes and line it up with some vertical line in the background—say, a door jamb.

Now, close your left eye and observe whether the pencil has moved away from the jamb-line. If it has not moved, then your open eye (in this case the right one) is your Master Eye.

If the pencil should move upon closing your left eye, open both eyes and close

ILLUSTRATION VI.

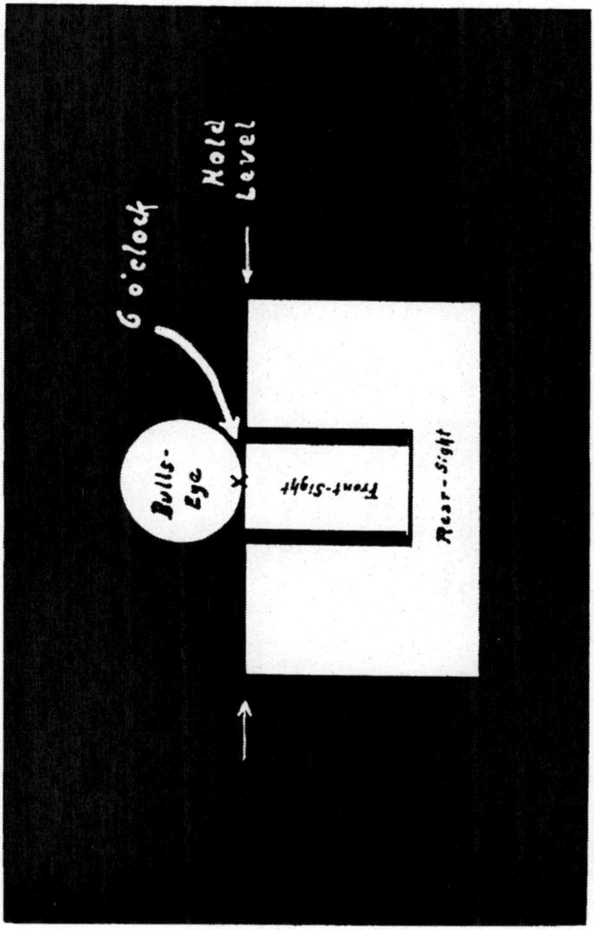

ILLUSTRATION VII.

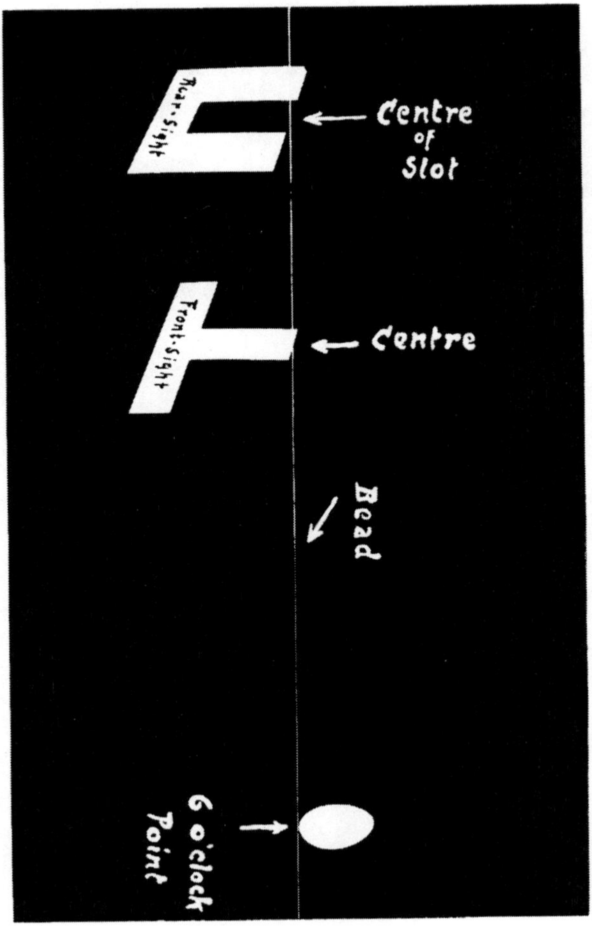

the right eye. You will then find that again the pencil remains stationary and (it being open) the left eye is your Master Eye.

After having ascertained your Master Eye, use it for drawing the bead, but keep the other eye open too.

Look straight **at** the target, and **not through** the sights. The idea is to have the viewing lines from both of your eyes converge upon the target. The angle will be less obtuse than if you had the lines converge upon the rear sight, and, consequently, much less strain is encountered, as your eyes need to go into one focus only.

All this is really much less complicated than appears on paper. Many pupils can do it at once.

The thing may need a little mental piloting on your part. Get yourself to concentrate both your eyes solely upon the target and let the sights fall in line behind unconsciously.

Come on, try this: Look at the target with both eyes open. Then bring your gun arm up slowly, with your eyes still trained on the target and **on nothing but** the target, until the gun appears within your vision.

Although you do **not** pay **any** attention at all to the gun and with both eyes glued to your target, you will have a hard time to avoid clearly seeing the barrel and sights.

The following birds-eye sketches will illustrate the why and wherefore.

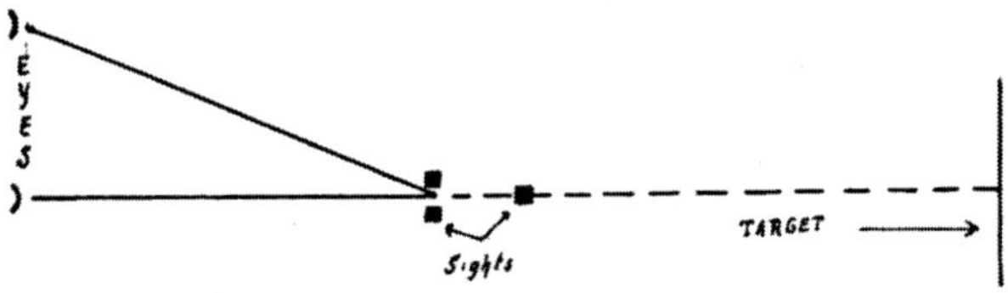

The lines converge upon the **rear-sight,** are then squeezed together into one line, which strained line then travels over the front-sight towards the target.

The effect is an unclear view of the target. You are sight-conscious and your poor eyes are being kept busy changing their focus all the time.

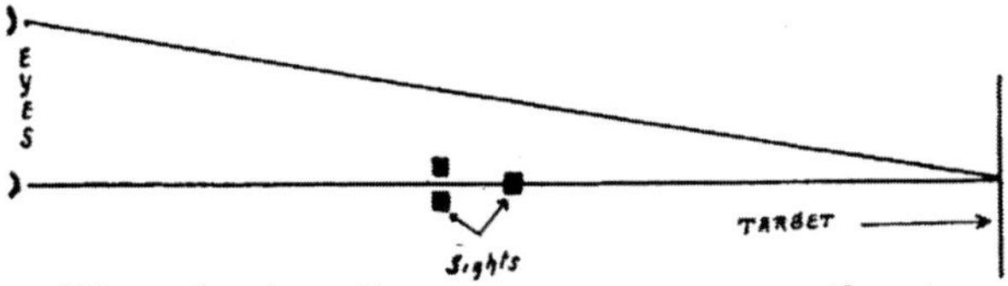

The viewing lines converge upon the **target** and the sights are lined up behind in an effortless, secondary operation. It is obvious that the effect is a plastic or stereoscopic view of the target—you are "target" conscious.

At first, you may see two images of your gun —one distinct and the other less pronounced.

After a short while, the secondary image will dim and finally disappear.

Needless to say that if your eye-sight is faulty, the oculist or the optometrist should be consulted, preferably a physician who is a shooter himself or one who takes an active interest in your particular problem.

We shall now assume, for a moment, that you are satisfied for the beginning with hitting the black only. (The advanced state of hitting the **center** of the black most of the time will come a little later.)

Are you satisfied to hit the black only?

Very well. We know that if your gun is sighted correctly, you should aim at 6 o'clock, and the shot — everything being right — should hit the center of the ten-ring.

Well, I say: DON'T AIM AT 6 O'CLOCK. DON'T.

Here, my friend, is a chance to show me that you are really intelligent. There are more important things than aiming at 6 o'clock. Really, there are.

You remember Grip, Stance, Relaxation — in the order of their importance. We have other important items, other more important things than sighting at 6 o'clock.

If trigger is squeezed while Frontsight waives within shaded area, a Bullseye will be scored.

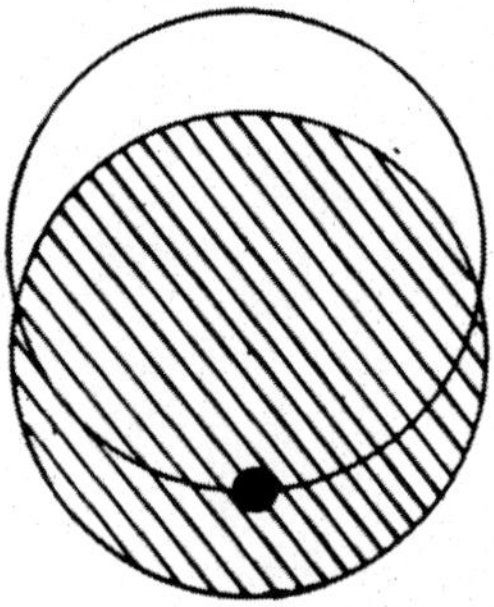

Upper circle represents Bullseye. *6 o'clock-point.

For the time being, I say again:

Don't try to hold your aim at 6 o'clock!

Your hand, and therefore the muzzle of your gun, are in constant motion. You cannot prevent this. If a person claims that he is able to hold his gun "as steady as a rock," he is just talking. I say, in contradiction ,the muzzle of your gun waives constantly left and right and up and down. With a healthy person, or a person of a stoical or apathetic temperament, these movements are less pronounced and less jerky. Still, the uncontrollable motion of the muzzle persists, and should be acccepted as normal.

Now, the bullseye is a good three inches in diameter. Allow your muzzle to wander in or under this bullseye, (see Illustration on page 41) and squeeze your trigger slowly. Without anticipation on your part, the bullet will leave the barrel and drill a hole in or near the bullseye.

You will have to take my word for it until you try it yourself. I promise you that I shall prove right, if you will follow just one other little hint — namely, to

SQUEEZE.

I appeal again to your intelligence. Do you really believe that your mind is receptive enough to soak up wisdom without back talk? Well, then, for Pete's sake — SQUEEZE.

How you should do it? It is very simple.

Slip your index finger into the trigger guard, not too far, not too little — just handy and comfortable.

Then find the trigger.

Barely touch it—very lightly, as if it were something hot.

After you have established this zephyr-like contact with your trigger finger, say to yourself:

> Now I am going to crook my finger a little — just enough to let the trigger depress the skin.

Now I am going to bend my finger 1000th of an inch — now another 1000th—

(Let the muzzle wander wherever it will, so long as it sticks somewhere near the bullseye.) (See Illustration page 41.)

Now another 1000th —
Hold it!

It seems that we are going a little too fast.

All right! We'll now try half the speed. Hold it all the while. Don't let go of what you have.

Another tiny impulse. B I N G !

Bullseye!

Oh, oh, yes sir, a bullseye! Some kind of a bullseye, but a bullseye just the same.

And you didn't even know when the shot went off.

This, my friend, is what I mean by Squeezing. Treat the trigger as if it were brittle glass.

You remember that you crooked your trigger-finger 1000th of an inch at a time — then you **slowed down** by half. Not bad for a start.

It is not necessary to take a micrometer to measure the movement. Just get your imagination to work. Visualize a situation where a dear (?) relative who has slightly overstayed her visit, according to your notion, is approaching with the intention of announcing that she will stay permanently. You can already see the purposeful glint in her eyes, but we assume that she knows that she dare not talk until you have finished squeezing. (If you consider the slow squeeze an ordeal, just think what tortures she must be undergoing, you heartless brute!) Pretend that the slower you squeeze, the longer it takes her to get at you.

Are you listening? Please wipe that expectant intensity off your face. (This is after all only a psychological experiment, and your gun is empty and not pointed at any one in particular.)

If you can bring yourself to cut that really slow speed in half again, you are still going too fast. Get the idea?

When I said that you should crook your finger, I was speaking in general terms. If I were fussy, I should have said (and I am saying it now):

> Don't crook all of the trigger finger. Just crook that part of it lightly which touches the trigger. Don't exert or bring into play any major muscle — certainly no muscle of your hand, wrist, or arm.

Muscles don't do your shooting at all. Their function is to hold the gun up as lazily as possible.

This delicate treatment of your trigger is an important key to success. Again you will have to take my word for it. Don't argue. Just try it. And, I dare say, any person who cannot make himself do with the trigger-finger what I have advised, is not well controlled enough to become an expert shot, and he might as well start practising with a trench mortar or a snow shovel.

BREATHING

This, although important, is very simple to do.(True, you know how to breathe, after a little spanking, upon your entrance into this vale of tears. However, I am talking of conscious and measured breathing, suitable for shooting.)

Now, without exertion, inhale and exhale slowly and deeply a few times.

When you are ready for your shot, inhale normally. Let some of the air out of your lungs, and then lock your throat.

With a comfortable amount of air locked in, none of your chest and neck muscles are strained, and you can easily hold your breath (which you **must** do) while you take aim and fire.

Do not hold your breath so long that your hand shakes while aiming. As soon as you begin to feel that the gun is putting on weight, lay it back into your left hand and put your left thumb under the cocked hammer.

When you are again ready, fit the gun back into your gun hand (See Illustration I) and free the hammer only after the fitting is completed. That way, a cocked gun will be safe at all times.

LET'S WATCH

In target shooting you must actually **do** certain things. (Paradoxically, positive **inaction** seems to be one of the things.) Just wishing to do these things will not do.

To show what I mean by actually doing, let us for a moment marvel at what takes place within the short time that elapses between taking aim and hitting the bullseye.

The body of the expert shows in perfect

stance, with raised gun. (Illustration V.) Every muscle is relaxed, from the little toe to the trigger-finger. The facial expression is one of profound composure. Both eyes are open to target and sight. (The supreme effort of the expert is towards the achievement of this relaxed poise — through perfect coordination.)

A picture of absolute and tenseless inaction! (Illustration V.)

Now, an imperceptible caress of the trigger (being merely the thought of a motion).

A Crash! And a hole in the bullseye!

There is none of the flash and elan of sabre blades or the graceful movements which one sees in a foil exhibition.

Apparently there is a lack of the dramatic altogether.

Yet, can you conceive of anything more startling or awesome than the deadly lightning that flashes from out of that statue of studied relaxation—smashing out the center of the Elusive Ten?

RESOLUTION

Concluding this first series of chapters, I am almost tempted to say that constant practice makes a master. This may sound a little trite, but so do lots of things that are very important truisms.

We enthusiasts take our shooting very seriously. We do not want to stay "just average shots." We want to get somewhere. That is why we practice daily.

It may be that all of us do not have the facilities for shooting on a range every day. However, there is the great expedient of "Dry Practice."

Take six empty shells and fill them with lead. Use these weighted shells in your cylinder and snap away. It will do no harm to a good gun, and it simulates sharp-shooting well enough to be very valuable.

Spend a few minutes each day in dry practice, aiming at a little spot on the wall and you will be surprised at the speed of your progress.

But, do not waste any shots, even though you are only snapping. Take aim as carefully, call your shots as painstakingly, and be as honest with yourself as a real target would be.

Make a firm resolution now to practice every day — and stick to it.

Keep a constant watch on yourself, particularly in regard to coordinating the functions as practiced singly by you.

Just squeezing the trigger will not score, if you forget about the grip and the other items dealt with.

Among the seven functions, (see page 17) some may be less important than others, but they all are cogs which fit definitely into the scheme.

Resolve to check up every time you shoot to determine whether or not you are forgetting one of the seven functions, and if so, correct the omission.

PART II.

READY FOR THE EXPERT

As you progress and finally emerge from the rank of "dub" into the loftier regions where you are reasonably sure never to drop your scores below 75 on the Standard American Slow-Fire Target, you are ready to get acquainted with the finer points of the game which have been evolved from practice and study.

And right here, I wish to point out a threatening barrier to those, who, after having read this far, would try to practice what follows here, without taking the trouble of first becoming proficient on the strength of the preceding chapters. Remember the seven fundamentals? (See page 17.) Have you absorbed them? Do you now effortlessly, systematically, follow them? Otherwise, following the finer points will only hinder you in the first steps of your development —by their very number and elaborateness.

You must understand that the method propounded by me, rests largely upon suggestive evolution, if I may call it that. The system leaves you free to concentrate on fundamentals, along simple lines, such concentration making for quicker progress, as you go along to work in stages.

Only when a certain stage of perfection has been achieved should the finer points be assimilated.

You see, these pointers seem sometimes in contradiction to what went on before. They only

seem to be; however, in reality, they are an elaboration.

If you were to try to swallow your dinner in one gulp, you would get indigestion, but after you have learned table manners and really know how to appreciate a gastronomical treat, only then do you fully appraise the finishing touches which the gourmet applies in the shape of an old cognac and a fine cigar.

FINE POINTS

The most important of the fine points seems worthy of a lengthier treatment.

It is the matter of squeeze and grip as a combined unity.

Intentionally, the grip was treated rather loosely in the first chapters. The squeeze, although approached with greater elaboration, is not complete unless a direct and thorough unification of both squeeze and grip has been accomplished.

As this treatise deals mainly with the revolver and not with the automatic, the unscientific construction of the revolver must be recognized and, perforce, be overcome by several muscular applications.

The metacentre of the revolver is wrong, we should say, if there were an exact science of revolver-shooting. (Maybe there is, but we needn't bother). Such centre would be found to be composed of the relation of centre of weight to the mysterious action of the recoil, superimposed by the mechanical action of levers, springs, friction, etc. Then there is the fact that the centre of weight changes with every shot. The weight-load of the cylinder gets lighter after every cartridge fired and thus influences the action-coefficient.

I know this sounds semi-scientific. However, I defy anyone to prove that the revolver is really the answer to the prayer, by the expert, for the ideal target gun.

As I said, the inherent weaknesses of the weapon will have to be counteracted by certain muscular applications. (The curious fact that all major target records were made with revolvers is merely an example of the statement that its weaknesses can be overcome through correct muscular applications. Still, the automatic, in my humble opinion, is going to take the place of the revolver in time to come.)

The directional actions of the trigger and hammer, to begin with, must be counteracted.

The backward squeeze on the trigger tends to pull the muzzle down, while the fall of the hammer will jar the weapon out of countenance. We can now, by using a downward hold with the thumb on the cylinder-latch in combination with the position of the ball of the thumb, practically counter-balance the directional actions mentioned. If we support the gun further with a moderate side pressure of the lower part of the little finger and of that part of the trigger-finger which touches the frame, we further minimize the evil forces.

The test is simple. After cocking your gun, place a ten-cent piece on the top of the round barrel near the muzzle, behind the front-sight and squeeze. The coin should stay on the barrel. It is a matter of a little experimenting to find the right minimum amount of thumb and finger pressures necessary to prevent the coin from dropping off.

The proper squeeze, combined with the judicious use of thumb and little trigger-fingers for support of the gun action, has, you will remember, been mentioned in the first part.

I also advocated in a former chapter the absolute relaxation of the muscles of the shooting hand, although I spoke of opposing pressures. We must now readjust ourselves by accepting consciously a few minor muscular applications.

Mind you, I do not say even now that you must religiously do thus and so. I am simply again pointing out certain places where the gun should be supported. It is unique with every one how to achieve this support. If all our hands were alike and if we all were using the same gun with the same action, I might attempt an exhaustive elaboration. What should happen, as you know, is this:

The **forepart** of the trigger-finger applies a delicate, zephyr-like action on the trigger.

The **lower** portion of the trigger-finger exerts a fine steadying pressure against the frame of the gun.

The thumb, resting on the cylinder latch, steadies the trigger action effect on the gun.

The ball of the thumb touches the back strap.

The middle finger supports the main weight of the gun and also acts as a sort of fulcrum in cocking.

The ring and little fingers complete the grip (although the ring finger does not call for any muscular action).

While the tips of middle, ring, and little fingers must not touch the stock, the little finger where it touches the stock with its **lower** part, should exert a light steadying pressure to the left, illustration II, VIII.

There is no perceptible bend at the wrist. The whole arm, wrist, and hand point effortlessly forward. You find now, to your surprise, that the matter of grip, all at once, has become of major importance, and rightly so. If you now combine squeeze and perfect grip, you certainly do have a fine equipment.

Still, my friend, remember again: The main part of the hand should not be tensed; neither should wrist or forearm muscles.

To impress upon you further the dominant importance of an individual and correct grip, I make the unorthodox statement that such a grip is, indeed, the crux of the art of revolver shooting. Things like the squeeze recede into insignificance when confronted with the question of grip.

To illustrate, let us again take under cursory observation the actual mechanics of Squeeze. What is the difference between a **pull** and a **squeeze?** There is none. A **squeeze** could be classified as a modified **pull.**

We teach the squeeze in order to get as little mechanical disturbance on the hold as possible. While squeezing, the shooter necessarily tends to maintain a uniform hold much better than if he were just pulling thoughtlessly.

A squeeze, according to this conception, would be an action which lets the muzzle dip or

ILLUSTRATION VIII

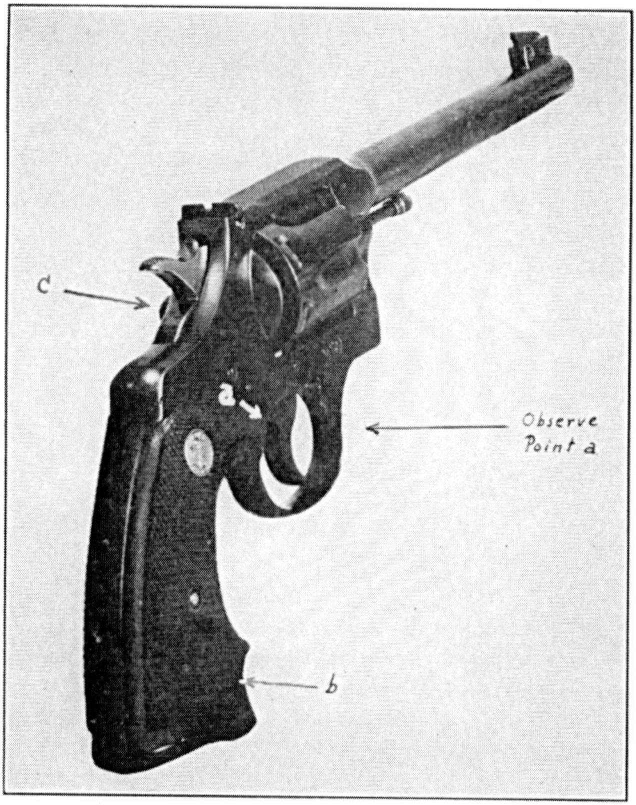

Point "a" on frame against which the lower portion of the trigger-finger exerts a slight steadying pressure.
Point "b" against which the lower portion of the little finger exerts a slight steadying pressure to the left.
Point "c" is the cylinder latch on which the thumb is placed.
The picture permits a view of the special inset which lowers the position of the middle finger. The stock has been lengthened correspondingly by the same insert, which is in one piece and can be easily detached.
Also the back-strap against which the ball of the thumb bears, can be clearly seen.
When looking at the Rear-and Front-sights one should also study the illustrations VI and VII.

tilt as little as possible. The latter result, however, is really not directly achieved through the perfect squeeze, but through the improved holding conditions which we shall be able to maintain by the deliberation of the squeezing action. The secret, therefore, is to find a hold which is easy to maintain and so applied as to overcome the more or less pronounced disturbance caused even by squeezing. I venture to say that it would be entirely possible to actually **pull** the trigger and hit things, if one assumed a hold which would wholly counteract a given violence of such pull. Doubtlessly, it is easier and more natural to use a restrictive hold for a smooth squeeze than for a thoughtless pull. That is why we favor the **squeeze** and condemn the **pull.**

Squeeze the trigger gently and slowly and you will be in an excellent position to find and maintain the ideal grip or hold on your gun. A squeeze alone will not perform unless it has as a basis a suitable grip. Hence the importance of grip. I urge you to concentrate all your faculties on the task of finding your individual grip on your gun. I stressed before, on pages 25 and 26, that your hold on the gun should entail no appreciable muscular effort. It is agreed that to support the weight of the gun, the middle finger will not be required to be active at all. Realizing this, there are really only two functions of the gun at the time of discharge which demand attention.

We have (1) the actions of the trigger and hammer before the explosion and (2) a certain twisting effect made possible by the recoil. The first tend to pull the muzzle downwards and then partly upwards again, while the latter may throw the muzzle sideways if the hold is incorrect. What causes these movements and how can they be counteracted?

Let us consider the first series of motions: While you depress the trigger you work against a spring and lever arrangement inside the gun. This resistance, with the middle finger as a point of fulcrum, will naturally cause the front end of the gun to tilt downwards. The actual click of the hammer will, on the other hand, throw the muzzle slightly upwards again. The ball of your thumb resting against the back-strap, in juxtaposition to the middle finger, is fully able to offset the downward motion, while the thumb itself, with a slight but steady downward hold on the cylinder-latch, neutralizes the upward flip of the muzzle caused by the fall of the hammer.

We come now to the side-twisting in connection with the recoil. The recoil really should throw the muzzle only straight upwards if it were not for the inefficient way in which the novice holds his gun. (The theoretically straight upward jerk caused by recoil is not sought to be corrected since it has been compensated by the height of the front-sight.) Usually, the novice clamps the tips of his middle, ring, and little fingers against the stock. This, of course, will tend to direct the recoil-motivated muzzle to the left. To correct this, the poor novice is taught to press with his thumb sharply sidewise against the frame. He is expected to actually work one side of his hand against the other instead of latently neutralizing expected forces by strategic placement of support.

Now, suppose you take my advice and keep the finger tips away from the stock. You will find that the recoil will be upwards only. Refraining from further theorizing, let me again make the statement that a revolver should be supported on five points:

(1) The function of the middle finger as a point of fulcrum and to carry the weight, is self-explanatory.

(2) The ball of the thumb bears against the back-strap.

(3) The thumb acts downwards on the cylinder-latch.

(4) The trigger-finger, where it touches the frame, exerts a steadying side-wise pressure and so does

(5) The little finger, with its lower portion. (See Illustrations II and VIII.)

How can we, with these various strategic applications, maintain a tenseless grip? Let me advise you again. I shall use a little example first. If you open your right hand and place a playing card flat against the palm, it is possible for you, by hooking one edge of the card in the loose skin of the lower portion of your middle finger and the other edge of the card in the ball of your thumb, to hold the card in your now relaxed hand without any muscular effort at all. The skin tension between the two points of contact (middle finger and ball of thumb) is sufficient to clamp the card securely. We do the same with our gun. We clamp the stock in the palm of our hand by first stretching the palm, and, after we have hooked the front and back ends of the stock at the two points mentioned above, we just relax the hand. All we do then is to place our thumb and the other fingers. The gun now is held quite effortlessly, and the tremors caused by rigid tension will be absent.

We shall have to experiment to find out how far the middle finger should be pushed under the trigger-guard. If it goes in too far, the gun will shoot too much to the right. By shifting the hand forward or backward, you will soon find a position which permits the recoil to throw the muzzle straight upwards. And this position, once found, should be **memorized** and assumed **every time** you take the gun in your hand. Once you find the position you are over the hill and on your way to the expert medal.

Mention should be made of certain media which, experience has developed, will facilitate the search for the ideal hold. With few exceptions, the shape of the grip of the revolver is such that, with the middle finger supporting the weight of the gun, the trigger-finger does not lie horizontally. Indeed, the trigger finger is pointing very much downwards. Instead of just being crooked naturally when squeezing the trigger, the finger is forced to exercise the backward pull on the trigger in a way which is foreign to its anatomy. The trigger is easiest to work when pulled straight backwards. The trigger finger in its unnatural position will, perforce, pull upwards too. This fact has been recognized by experienced shooters, if not by some revolver manufacturers, and several remedies have been suggested. There are on the market certain insets which can be attached to the stock. These attachments will place the middle finger lower and will, consequently, make the trigger finger assume a more horizontal and natural position. These attachments, however, have the common fault of shortening the available stock to such an extent that many shooters will not be able to place their little finger sidewise against the stock. I overcame this deficiency by having my dentist make a wax impression of my hand grip on the gun and then cast this impression in hard rubber. With my individual attachment, the cost of which was exactly $5.00, I now have a stock which is amply long for my hand and, at the same time, my attachment gives me a very comfortable feel. See Illustrations I, II and VIII.)

I find that this dissertation on grip has become rather involved and penetrating. I fear I may have overstepped the frame set for the Manual, but I also believe that I could not have treated a difficult and important subject in a cursory way.

The sighting is, of course, now synchronized with the "squeeze".

It does not do any more to waver around the bullseye as advised in the first part. (See page 41.)

We must now aim patiently at 6 o'clock. (Illustrations VI and VII.) Since we cannot hope to "keep" the bead at 6 o'clock while we are going through the slow process of squeezing, we must compromise. Every time the bead passes below 6 o'clock, we squeeze. When the bead has passed by, we stop squeezing but hold what we have until 6 o'clock shows favorably again, when we continue squeezing — and we hold and squeeze until the bullet emerges.

The beginner will hastily **pull** the trigger as soon as he beads at 6 o'clock.

(That is why I advocated in the opening chapters to sight not at 6 o'clock but at an imaginary circle — (See page 41) just so that the squeeze had a chance to be developed).

Our object now, of course, is much more intensified. We are not satisfied any more with just hitting the black somewhere. We are now after the "Ten Ring", the centre of the bullseye. (The world record is 10 consecutive hits in the Ten-Ring. This accomplishment is called a "possible" and the rating equals 100.) It will be found now that the "Ten Ring" proves rather elusive. This tact, in addition to being very aggravating, virtually suggested the title to this Manual.

At this stadium, you may expect some help from your physical system. Through varied practice, there has developed within you a hitherto latent faculty—namely, the ability to observe a variation in the fine vibrations of your system, as seen at the muzzle. We have acknowledged the fact that these vibrations are inherent with the human body, coming from heart-beat combined with certain nervous pulsations, tremors, etc.

It is logical to believe that if such vibrations are accellerated through physical or mental exertions, the muzzle will respond in accordance. On the other hand, a complete lack of exertion would permit the muzzle to register only the normal pulsations. Now, if we are completely absorbed in the task of sighting and squeezing (everything else being right, and you should see to that), we will observe distinctly a variation in the pulsations visible at the muzzle. They seem to slow down at times, and there occurs an infinitisemely short period where the pulsations are so diminished that the muzzle does not register them visibly. This moment is not of long duration and if it can be co-ordinated with your mechanical task of sighting and squeezing, the resulting shot will emerge under ideal conditions.

I have heard it seriously propounded that all a shooter can hope for, is to endeavor to stay in the black. If you were to resign yourself to such an attitude, you could score good targets only if Lady Luck were at your shoulder. It is entirely possible to stay in the black and score only an "80." If you are lucky, a few shots may slip into the "9" or even the "10." (We shall dismiss such an attitude forthwith, since we are supposed to have graduated from the beginner's course in part I.) What kind of unintelligent shooting would that be? I had much rather see a **close group** of shots, even if such group should not happen to be all in the black. If you are able to shoot close groups (and that should be your aim), you are in a position to shift them by correcting your hold, sights, or stance. You are, at least, doing conscientious shooting instead of having to trust to luck all the time.

Before this chapter's conclusion, I wish to relate an amazing experience. I know one of the instructors of a public institution who is considered quite a shooter and, who, quite possibly, landed his position on the strength of his achievements on the range. A most appalling school of thought found in him a vigorous proponent. The guiding idea seemed to be that since a beginner will always **pull** the trigger and since pulling will place the shot wildly below the paper of the target, he should be made to hold at the upper edge of the paper target. This, he said, would pull his shots somewhere near the black. Then, as the idea of **squeeze** is slowly hammered into the poor pupil's cranium, the bead is gradually lowered until the pupil can safely approach the 6 o'clock-hold. This process appears so unintelligent to me that comments are superfluous. The only merit of it seems to be that, even if disastrous, it is infinitely slow, and if interrupted in time (before the pupil has lost all interest) it might not wreck the pupil's chances for all time.

We have here one of the reasons why the idea that the safest place is in front of an officer's gun is so prevalent. I have cited only one example, but it should be illustrative.

TIME-AND RAPID-FIRING

I believe that both kinds of firing are easily acquired if they are approached in the right manner. However, before one starts to take an active interest in these two kinds of firing, I strongly advise (and I know what I am talking about) acquiring a certain proficiency in **slow** fire first. I consider it positively assinine to rush into things if a solid basis is lacking.

Don't forget, slow fire is the foundation of everything. In slow fire all the rudimentary elements of shooting are covered and, only after a shooter has reached a certain goal in this kind of firing (whether the goal be set very high or not), will he be able to approach Time-and Rapid-firing with any semblance of promise.

By some quirk, certain pupils seem to make better scores in rapid firing than in slow. (A possible explanation is "a frantic clutch on the gun and a tensed-up condition of the mind" —

things which cannot be maintained for any reasonable length of time.) Their efforts are sporadic and their attitude is as much the fault of the coach as of themselves. Unless a radical resolution changes things from the ground up, these shooters are hopeless. They are facing what the French call a **cul de sac.** They are spoiled for slow and time-firing, because they are violating all the elementary requirements . . . and as rapid-fire shooters they have no consistency. In short, as target-shooters they have no future. A good slow-fire shot, if he is interested enough, can acquire proficiency in all kinds of firing because he has the proper foundation. A rapid-fire flash is nothing but a flash in the pan.

There are certain requisites which should be mastered before one attempts to score on the range. Among them "time-rhythm" seems to be the most important one. To be specific, one should be able to fire five shots with **equal** intervals within a given time period, without resort to the stop-watch. Take, for instance, rapid-fire where the time limit is 10 seconds for every five shots:

Raise your revolver, (This practice should be done with practice shells) aim, and count slowly:

"Twenty-one and"

"Twenty-two and"

(Each count to cover two seconds) and you will find that at the end of . . . . . . "Twenty-five and," a period of ten seconds has elapsed.

It would, theoretically, be advisable to start shooting with the whistle. However, such practice tends to key the shooter up too much. It is much safer not to start shooting with the blast of the whistle, but to consider the blast only as an announcement, rather than as a warning signal.

When the starting-whistle blows, calmly start counting "Twenty-one and" and try to have the hammer click every time you get to the chosen syllable "-one" "-two", etc., like this: (I shall emphasize the syllable)

Twenty-**one** and
Twenty-**two** and
Twenty-**three** and
Twenty-**four** and
Twenty-**five** and.

Get in the habit of firing always at the same syllable.

After you are able to follow the precept automatically, in dry-shooting (without getting flustered any more), you may be considered to have mastered the "time-rhythm" and you are ready for real fire practice.

As this dry practice can be exercised anywhere and any time, the serious pupil will endeavor to keep at it in idle moments. The progress will be astonishing, especially if such practice be adhered to every day, if only for a few minutes.

In time you will not need to count anymore and the time-rhythm will be followed almost mechanically, without effort on your part.

Thus, since no effort is wasted on the timing, and since the rapid cocking of the hammer has, meanwhile, become second nature, your full attention can be centered upon aiming and squeezing.

Even the aiming, by and by, will become automatic, and the only action you cannot mechanize, namely, the squeezing, can be attended to with perfect leisure.

Upon going into practice with sharp shells, it will be found that the recoil (which so far you have more or less disregarded) comes to the fore, and the necessity of maintaining a more pronounced grip becomes more apparent. Your mental and muscular reactions must be studied with a view to finding a grip which will cause the gun, after every shot, to find itself in exactly the same position as before the shot without having to grope for the original grip. The recoil throws the muzzle up, and it has to be brought down again quickly towards the aiming point in a short straight line. However, you have to **cock** the gun before it can be brought back to the target and this cocking should be done **without** losing your predetermined rapid-firing grip. In fact, the whole secret of the new grip is that you should not only be able to cock quickly, but it should also embrace all the favorable elements of the slow-fire grip, so that you can get the proper supporting and squeezing actions for accuracy's sake. Use the recoil to facilitate rapid cooking, but **find** the universal grip. Constant practice and watchful experimenting cannot be dispensed with. One will find that the lighter recoil of the smaller calibres offers less of a problem than the heavier ones.

There is a grip to fit any calibre, but it is entirely up to the shooter alone to find his own.

As for breathing, it is recommended that you h o l d your breath for the entire ten seconds when rapid-firing.

In time-firing, where we are allowed twenty seconds, take a fresh breath for every shot.

In order to be able to hold your breath for the period of ten seconds, it will, of course, be necessary, to take in a sufficient amount of air before starting to shoot. However, do not overdo this. Practice will show you the minimum of air you can safely inhale and hold without discomfort.

All in all, in time and rapid-firing, the method of shooting is not very different from that of slow-fire, as described. It may be then that the squeeze is, of necessity, speeded up a little.

But, let me assure you that the moment you start tensing up or p u l l i n g the trigger, your targets are not going to be worth looking at.

In conclusion, I think, I should put a rather unorthodox question: Of what use really is "Rapid-Firing"? I do not know whether the question has ever been asked. So far, I have heard no arguments against "Rapid-Firing", but I wonder.

Suppose, we take the case of a police officer who is engaged in a gun duel with a gangster. (Such a situation would surely be quoted in support of rapid-firing). According to present precepts, the officer is expected to start rapid-firing. It is logical to assume that in this kind of fast firing accuracy must suffer. The chances are that, coupled with the resultant excitement, the accuracy of the officer's shooting is further greatly reduced and that the officer misses his antagonist entirely. After the last shot has been fired, the officer's gun will be empty and he is robbed of his most potent weapon until he finds time to reload again. The officer knows that it is necessary to hit his opponent only once in order to demoralize him. But he has to be **hit**. The value of noisy display is questionable, but a direct hit is a powerful persuader.

Is it not then logical to say that the problem seems to be more one of "getting the drop" and, if necessary, to "hit with the first shot" and then to be ready for further emergency?

If the officer's training were conducted along these latter lines, the lawless element would entertain the idea of a pistol duel with an officer with a much greater amount of trepidation.

Since, however, Target-Rapid-Firing in its present form is an established institution, we must submit in order to be able to enter target competition at all.

My suggestion for competition, if I were asked about it, would be to replace the present form of rapid fire with the following:

Gun in holster.

Upon command the gun should be drawn and fired **once,** the shot to hit a rather small target. The gun should then be re-cocked quickly and held up again at "Ready Fire" position, aimed at the same target. The whole action should be timed and the results be judged by "time elapsed" and "value of hit," the weight of scoring to lie with "time elapsed."

ADJUSTED STANCE

It happens sometimes that the group of your shots appears elongated horizontally. (If this distortion is caused by body sway (see First Chapter under "Stance"), then we have an effective remedy.)

By advancing your rear foot slightly you change the sway of your body. Instead of swaying from left to right with your gun-arm, you will now sway more up and down, thus causing the group to constrict horizontally.

If the group should appear elongated vertically, follow the opposite precept.

Remember, if you present more of your shoulder to the target, your gun arm will sway from left to right. If you present more of your chest to the target, the arm will sway up and down.

FLINCHING

Aha! you will say, here is, maybe, the spot where I can learn something worthwhile. Well, this Manual deals with practical shooting and not with psychological problems.

Your flinching, my friend, is apparently something very undesirable. It might spoil your aim or prevent you from calling your shots.

Do you flinch because you like to? Or just on general principles?

Well then — for Pete's sake —

It is really unimportant, though. However, you may differ with me. Maybe your method of flinching is not of the crude kind that we other earthworms were afflicted with.

Maybe you would cultivate it, and do your flinching so perfectly that you might feel inclined to aspire for higher honors.

Personally, I again proclaim this topic unimportant. I find no time, when shooting, to waste any thoughts upon flinching. All my attention is centered upon really important tasks, such as GRIP, SQUEEZE, SIGHTING, etc.

And, if I were you, I would not try to get rid of the **flinch**. Ignore the little imp, and if you work diligently on the important tasks, you will not have t i m e for him and you will, finally, concur with me that **flinch** is really unimportant.

You may, however, put only three sharp shells in your cylinder, while practicing, twirl it after every shot so that you do not know which chambers are loaded, and squeeze out six perfect actions.

Keep both your eyes open and see that your muzzle does **not** flick when you hear the empty hammer clicks only.

WOBBLING

A horrible word!

After a strenuous day at your business or office, or after muscular or mental exertions of any kind, a certain amount of wobbling may appear.

The characteristics are a more or less pronounced tremor at the muzzle, sometimes coupled with a surging up of nervous waves and a dimming of either the target or the sights.

In such cases, if at all possible, you should endeavor to lie down for a short time, seeking, first, complete **relaxation**. Follow up this period of rest with careful and conscientious breathing exercises.

After you feel that you have recovered sufficiently, you should start with practicing the one-two system of firing; that is, you fill every second chamber of the cylinder only, and fire

One Dry
One Sharp

etc.

The same method may be followed, if you should find yourself too tensely keyed-up.

This sometimes happens before a match and may be a sort of stage-fright. The empty clicks will reassure you and will also give you a chance to observe your gun muzzle, indicative of whether you pull or squeeze. Wobbling is successfully combatted if you follow the hints set forth in this chapter.

Another timely hint on this occasion would be: Never try to shoot with a full stomach.

Also the idea that one can shoot better after partaking of a stimulant — such as alcohol — is a fallacy.

It is not stimulation you need in shooting; only absolute relaxation!

SLUMPS AND INHIBITIONS

Every game has a bugaboo. In revolver shooting we have "slumps."

Our sport, being what it is, namely the outward expression of one's ability to utilize his control over his nervous system, it stands to reason that one cannot hope to be always in peak condition.

In this modern world, with its nervous tempo, an even and uneventful flow of affairs seems to be a thing of the past.

Our lives are subjected to more or less violent emotions which cause disturbances in the equilibrium of our nervous system.

Now, since good results in shooting can be had only when one's nerve-machinery is purring smoothly, it is conceivable that results will not be as gratifying if, through an emotion or a succession of them, the even beat of the machinery is disturbed.

With persons who have a stoical temperament, the surface of the sea of equanimity is not easily rippled by a gust of emotion. On others, however, (and they comprise the majority) slight, and ofttimes unimportant emotions will have the effect of a "storm upon the waters."

Some of us have an amazing elasticity which smooths the ripples in short order, without after-effects. Others (and, again, they are in the majority) lack this elasticity. These latter I wish to address particularly.

You, my friends, will probably experience periods of slumps in your performances, periods which, according to your individual set-up will be of short or long duration. I want you to view these slumps without alarm. I want you to keep up your confidence because slumps **pass away.**

As long as you cannot avoid emotional stress (or think you cannot), the control lever to your nervous system is in weak hands and your shooting results are unsatisfactory. On the other hand, grip the control lever firmly and the slump is over.

Does this suggest something to you?

Leaving out disturbances of a purely physical order, let me say that if we cannot avoid stresses, let us at least try to acquire elasticity. Not being a pathologist, I cannot advise you very well on the subject, unless it be that I offer a sort of thumb rule:

Make it a habit to forget things of an unpleasant nature as fast as you possibly can. Don't get on the range with a disturbed mind.

This needs training, of course. But your **will** to forget things which are unpleasant, is half the battle. So far, you have permitted such things to rid themselves. You may have wished for elasticity, but you have never seriously tried to **will** it firmly enough.

I said that slumps will pass and therefore they should not cause undue apprehension.

Once you have acquired elasticity, you not only will overcome slumps more rapidly, but you will be able to eliminate all sorts of inhibitions too.

I know fully well that in discussing things of a phychological nature I am treading on dangerous ground. Still, I take consolation from my conviction that even the most learned psychologists are sometimes at variance with each other. Although their dissertations may show a convincing clarity of logic, sometimes they are contradicted by the representatives of a different school equally convincing — and there we are.

What if I confine myself to a sort of home-learned gentry possibly kick about that? For instance: a shooter at home in his basement range shoots "85" and better all the time. He has reached, as I call it, a level of 85. When he is requested to step up and repeat the score in public, he flunks. Here we have the same man, the same gun, ammunition, target, and conditions. What is it that prevents him from performing uniformly? It is true it is the same person, but upon the system of this person has suddenly settled the cloud of an inhibition. This inhibition whispers:

> "Now you are not alone. You are being watched by other people, important people whose opinion you value. For God's sake, don't make a fool out of yourself now. Don't dare to pull your shots or permit any to get out of the black. Why, that would be disastrous. Get a grip on yourself. Aim carefully. Don't let your hand wave like that.

Man, you are going to muff that target, I can see it. There goes the shot. By golly, it is a nine. How did that happen? I saw you close your eyes when the shot went off. Well, thank God, it is a nine anyway. But, be more careful with the next one. Come now, shoot nicely, everybody is watching. There, damn it, you pulled that one. There goes your 85. It's no use trying, you'll never make it anymore. Trust to luck then and bang away. Oh! Oh! Hell! Never mind. That gun didn't feel right in the first place. I am sure it would have been better if you had used some of the other ammunition. What score? 72? What in h . . . ?"

That is what your inhibition whispers in your ear. It whispers while you should be having no time to listen, while you should be paying full attention to the mechanics of shooting.

Now, suppose you listen to what I have to say — and I shall not whisper. Don't endeavor to banish the inhibition by trying to do the opposite. Don't kill hot with cold. Out of a **positive** inhibition, don't make a **negative one.** Instead employ another, a different, positive habit.

To show what I mean, let me give you an example: Assume that you have had the bad habit of getting up at 10 a. m. on Sunday mornings (a positive habit). Don't foster the thought that from now on you are **not** going to get up at 10 a. m. again (a negative thought), but say instead: "From now on I will get up at 8 a. m." (which worthy resolution is again something positive). Get the idea?

Applied to shooting, don't say: "I am **not** going to listen to the voice of nervousness." Say instead: "I am going to think along the following lines when I am up front:

> "Well, here we are, old sock, old bean. Well, well, this does not look any different from my basement range. Why, conditions are better here. I don't have to be afraid of scaring the household with my banging. Nice people here, good friends all. Why, I might do right well. Just look how nervous some of the other shooters are. I wonder what about? Any one has his little troubles, I suppose. Come to think of it, I had very little trouble today. Well, well, if these birds persist in being nervous, they will be playing right into my hands. But, get set, boy, oh boy. It's your turn. Now, in the basement, the last time, I shot a 91. They expect me to shoot only 85. Maybe I can show them something. Remember: grip, squeeze, etc. Hell, I do that automatically. Here goes. Hm! Not so bad. This one, I jerked a little to the left. I saw it though. Called it 7 o'clock in the nine-ring. Let's look. Right. Another one! Right on the nose. Life is not so bad after all, eh? Well, fellows, there's your target. What's the score? 86? Fine."

A **positive** thought of that friendly nature is just as powerful over your nervous system as the sinister one. It does not scare you or make you tense, but it gives you the supreme deliberation which is the backbone of good shooting. I could go on proposing autosuggestive guides till I were blue in the face, but that is not necessary. As long as you have seen the general drift, you can make your own medicine — maybe better than I can. And that is all I am going to say on the subject for a while. Whew!!!

MATCHES

(Argumentum ad ignorantiam)

Let's have a match. There seems to be quite a controversy about the advisability of stàging matches with a view to getting shooters used to them.

The pros and contras sometimes wax quite heated. I am afraid I cannot entirely avoid the issue.

My own humble opinion is that a team or an individual should, through constant practice, first try to reach a **certain** stage of consistency, the limitation being, of course, set by the ambition of the team or individual.

A team, for instance, which averages in competition consistently 255 over the N.R.A. Standard American course, (Slow-Time-and Rapid) has a goodly chance of winning quite some honors.

Naturally, a team of that calibre has been through the fire, as it is obvious that it would average considerably higher if shooting under no competitive stress.

It really should be the ambition of every team or individual to reach this mark. The requirement to be met would be the ability to score consistently around 270 on the home-range. Actual competition will account for a drop in the scores.

Theoretically, it would therefore be advisable not to permit participation in matches, unless such minimum of 270 were first scored consistently.

Unfortunately, the eagerness of a promising pupil is often fostered by the coach and, against his better judgment, he permits competition on practically any score basis.

I believe that a shooter-in-the-making cannot learn anything at a match that will help him on his way up.

On the other hand, a match may do considerable damage. One must not forget that the shooter, in training for a match, sometimes for a period of weeks or months, must, of necessity, neglect his schedule.

I had much rather see a shooter advance along predetermined lines, curbing his impatience, **until** he has achieved his **set** minimum goal.

After that, matches against not too strong opponents (in the beginning), will be in order. In fact, from then on matches will be very beneficial.

I shall show you now the opposite side of the medal.

There are many, many teams and individuals who, without any concrete schooling, t umble from match to match.

They are consistently beaten by mediocre teams. Once in a while they may gain a shameful victory over an opponent who is even worse than they.

There is no honor in losing against an average opponent and surely none in winning from people who know less.

It seems as if a sort of madness had seized them. Match - match - match. That's all they seem to live for. Their scores get worse and worse. No sooner do they meet up with some club official than they start right away with

"Let's have a match" — as naturally as if they were asking a match for their cigar.

It is almost funny, if it were not so tragic.

My friends, the performance of an unfinished team is, necessarily, erratic under competitive stress.

In fact, even at the home-range the scores are unreliable. The very fact that the scores at home are not consistent should be sufficient warning.

I say, keep away from this craze.

Get good and ready first.

There is all the time in the world.

You do not have to be afraid of the stage-fright bugaboo. It is nothing but an argument to cover up a mania.

With the knowledge that you are prepared, nothing will frighten you. And, if you should lose then, you at least have lost to a strong opponent after a hard fight. But, with sound preparation you always have a fine chance to win.

Remember, your opponent is also under stress, particularly if he knows that you have come fully prepared.

REPUTATION

My friends, (I hope that I have, by now, earned the right to call you that) — there is one thing in revolver-shooting with which I have never become entirely reconciled. I should like to talk about this a little.

It is human for every one of us to desire to be a top-notcher, to be known as a marvelous shot,

—as a dangerous opponent, — as a champion — in short, to have a reputation.

If you should reach this stadium you have a perfect right to be proud of your achievement. Hearty congratulations are in order. You yourself know best what an amount of diligent work you had to go through to get this far.

You will come face to face with the realization by and by, that your work is not ended then—that you must go on and on. It is not possible for you to rest on your laurels. Remember, my friend, there is reputation not to "have," but to "maintain" — a very fickle thing indeed — and a danger.

Do not let this thing rob you of your peace of mind.

Do not let this thing grow into a curse.

I shall illustrate: You have a reputation. You are invited to a friendly shooting party — you go there expecting to have a good time — but you have to show your stuff first — and woe to you if you do not show in top form. Nobody considers whether or not you feel right or that conditions are different on every range.

Just because I was known to have done it, when in form, I was asked time and again to hit a cigarette at 60 feet with one shot with weapons I had never even seen before. I was too good-natured or too vain to refuse. I want you to believe me when I say that, although I succeeded, many a fine day was spoiled for me. Praise and admiration are not sufficient recompense for the silence which follows if you miss.

You are not permitted to experiment anymore. Every shot is expected to be perfect. You are under constant pressure and, sometimes, you wish that people would leave you alone.

The only way out seems to be to make light of your reputation. You may state that any average shooter could duplicate your shooting if he were to try seriously (which, we know, is true)— say anything — but don't let the thing get you.

YOU AND YOUR GUN

We have seen, after getting this far, that there is a little more to revolver shooting than many perhaps anticipate.

You don't just resolve some fine day to take up shooting, buy a gun, step up to the range and bang away.

You have found out by now that you have to get a foundation and that you have to work to get it, that you have to discard and to rebuild, to bend and submit to other judgment — until you can stand on your own.

Although everything is fundamentally simple, you have to make yourself "absorb" these simple things — and you have to practice, practice.

We have filled many pages with fundamentals and what will be considered as aids and hints, but, although we have criticized its technical weaknesses, we have not spent much breath upon the soul of our sport, namely, the revolver itself.

Do not, even for a moment, get the idea that your revolver is just a tool of the trade.

Every revolver has its peculiarities, although it may have been made with fine factory tolerances along with thousands of others — peculiarities, — indeed. I am tempted to say that every revolver has its own character. Don't smile, just switch guns and your face will lose its sanguine expression.

Let's pick a simile. Once you have chosen a gun, consider it in the light of a sweetheart, court it, fondle it, treat it with care; but, beyond all, **study** it, try to find its characteristics, solve its soul.

The rewards will be ample.

Your gun is very jealous and will not permit a rival (that is, not within eye-scratching distance). So, do not switch sweethearts (even though you may think that you have solved them individually) during a shoot.

Just try to do that very thing and your sunny disposition will suffer.

Therefore, don't.

THE POSSIBLE

It is timely that I should say now that within the scope of a Manual it is not possible to completely cover a subject in all its aspects. This would really require a volume. It is not even feasible. My aim was to show just the skeleton; to strip the difficult matter, as much as possible, of all finesse — unadorned with (for a beginner) unessential detail (Sight adjustments, choice of ammunition and a host of other things). I trust that this task has been covered in this handbook.

There may come a time when the advanced shooter, if he be ambitious, believes that he has reached his limitation. His hope is still that, one day, he will accomplish the "Possible" — but a sort of resignation seems to settle upon him. He may still take pleasure in knowing that he is considered a dangerous opponent, yet he recognizes that he is at a standstill.

However, a state of mind of that order is not at all in conformity with the hopeful enthusiasm which I have tried to instill into my reader — not at all.

Now, my friend, the battle has really only started.

If, so far, you have only followed suggestions, now is the time to start analyzing the cause of your apparent limitation.

Let us look the facts in the face. There is no reason in the world why you should not reach the "Possible" (10 shots in the Ten-Ring). I go even further and say that there is no real reason why you should not be able to shoot the "Possible" more than once.

The fathers of the Standard American Target were wily rascals. They settled upon the smallest possible Ten-Ring. Still "Possibles" have been shot. Luck does play a part in it, since even a machine-rest will not always shoot "Possibles" — with today's ammunition.

But, if you are seriously and conscientiously after the "Possible," you will come to shoot in the "nineties" pretty consistently. And why should you not have a little luck once in a while?

However — and here is the really important part — even if luck should turn her shoulder, you are still shooting in the high "nineties." Get the drift? With high nineties, you are a top-notcher. That can be and should be your goal. To get there, the word "limitation" should disappear from your vocabulary.

Use your head.

There must be some things which you are doing wrong, there **must** be. It is not the target, elusive as it may appear; it is not the gun. The ammunition, it is true, could be a whole lot more uniform.

But don't look outside of yourself.

You are the only one responsible.

It does not do for you simply to accept my statement; you should, inherently, discover it yourself.

What then is the purpose of this chapter? It is to make you realize that you are doing something which is wrong — what is it?

Find out! Study your problem. Experiment.

Why not re-read this Manual carefully. Some little thing may not have been absorbed. If, while experimenting, you are to discover for yourself an apparently more or less unimportant, yet very helpful expedient, and in glancing through this Manual you should find the subject already treated, would you not think yourself rather foolish?

Maybe, it is your grip, or the non-discovery of the infinitesimally short period where coordination brings about the disappearance of all tremor, just when six o'clock is easing on your front-sight.

Maybe, it is a number of things, and possibly only one single error.

Go ahead. You are on your own.

A complete coverage of all intricacies cannot possibly belong in the realm of a Manual. Consideration for the many who have not reached your status would forbid that. They must not be confused by chapters which really should go into a "Manual for Distinguished Experts" — maybe, sometime — later on.

Get your chin up —
And study —
And work —
To success.

FUN GALORE

We revolver addicts are a cheerful bunch as a whole, cheerful in the sense that we meet adverse happenings before the target with all sorts of alibis — alibis which sometimes are so extremely funny as to provide hearty amusement, —first for the other fellows and then (to a lesser degree) for ourselves. We go down, but we always come up again — for more. The bug has bitten us. We are addicts — hopelessly so.

Although afraid that our total sum of available amusement may be severely curtailed, I rather like the thought that your more or less sorrowful moments (which run concurrently with **our** bursts of humorous appreciation) should be lightened some way or other.

Take an acute case, a case to be found at any gathering of shooters. We find our friend exultant at the first signs of improvement. His bearing is full of exuberant vitality. He pats everybody on the back. His eyes glow. He struts around, and he is generally very happy. Life surely looks sweet to him. During the intervening week till the next gathering, everybody is his friend, even his next-door neighbor. His relatives do not pain him half as much. He is an amiable creature. And he confidently attacks his new target. But, for some reason, things seem to have changed. A cloud has passed over the sunny sky; friendship somehow does not seem quite as sweet.

What in the world — ? A pungent string of expletives indicates a frigid score on the target.

Something he ate the night before — the ammunition was faulty —not to mention the gun. That's it — the gun — no good — he knew it all along. He remembers having seen this or that fellow shoot some other model — that was the gun par excellence. All right, he'll get that other gun. His alibi is complete. He'll try again with the other gun. He's up again, bouncingly confident. Wait till he gets his hand on the other model.

And the cycle all over again — ad infinitum.

Brother, I wouldn't want to miss you among our bunch for anything. Life would be too drab otherwise. However, next time remember that there are ladies around.

Now, I know, you yourself didn't enjoy the show nearly as much as we others. I have a faint suspicion that you didn't enjoy part of it at all. You wish to avoid that in the future? Then: Don't make an ass out of yourself by fooling around with all kinds of guns, and fail every time. Stick to one gun — stay with it until you have it licked.

If you should get a stupid target.

don't alibi your way out

don't swear — and

don't blame your gun.

If you follow these "Don'ts" you will, probably, take half the fun out of our lives, but you'll enjoy yours just that much more.

If you are down, try — try again — with the same gun, and you will be up again, — and improve again. Because you are one of us. . . .

FINIS

It is with profound regret that I write "FINIS" to this Manual of mine.

I know that the subject has only been scratched. There is so much more to be said. But I promised a MANUAL and not a BOOK — and I have to leave much unwritten — things, which crystalized into concrete thought as I experimented and wrote.

Subconsciously, we are often doing things in certain ways. We do not become conscious of these ways until, by design or coincidence, we change them. We then find that results are not satisfactory either. We return to the old method which, however, upon analysis, makes us realize that we have been wallowing in the mud of mediocrity all the time.

And, if we are serious we **do** something about it. We think — sometimes we write. And, if the subject be artful enough, we go on seceding and deducing. And though, at times, we may believe that we have covered the fundamentals, the train of thought seems never to give out.

That is the way I feel about revolver-shooting.

Maybe, if this MANUAL finds favor with enough of you revolver-addicts, it might embolden me to "Go ahead and open the valve again."

Until then:

GOOD LUCK!

Smith & Wesson Hand Guns....................................McHenry & Roper
Hunting Pronghorn Antelope....................................Bert Popowski
The Woodchuck Hunter..Paul Estey
Small Bore Rifle Shooting.......................................Edward Crossman
Modern Bow Hunting..Hiram Grogan
Salt Water Fishing Is Easy.......................................Jerry Sylvester
Fishing Flies and Fly Tying.....................................William Blades
This Business of Exploring.............................Roy Chapman Andrews
Handbook for Shooters and Reloaders.......................P. O. Ackley
Greatest Fishing..Joe Brooks
Hunting Our Biggest Game.......................................Clyde Ormond
How To Hunt Deer...Edward Freeman
Burning Powder...Douglas B. Wesson
Practical Rifle Shooting..Walter Winans
Handloading Ammunition...J. R. Mattern
Shooting..J. Henry Fitzgerald
Rifles & Rifle Shooting...Charles Askins
Mauser Rifles and Pistols..W. H. B. Smith
The Book of the Pistol & Revolver...........................Hugh B. C. Pollard

Made in United States
North Haven, CT
04 November 2026

21772144R00055